Ricardo S. Barros

# Masonry Structures

Inspection, Modelling and Intervention

# Abstract

This study addresses the seismic reinforcement of masonry structures with the use of metallic elements. For a better understanding of the functioning of the metallic elements, it is necessary to understand the national and international panorama of seismic reinforcement, as well as the most important aspects related to these elements and whole process of seismic reinforcement.

There are various metallic elements, with several features that can be used taking into account their cost-effectiveness. Sometimes the application of metallic elements is expensive not only for its cost in itself, but also for the difficulty in implementation, which can increase costs. However, the use of metal elements allows for less intrusive interventions, which makes their use advisable when it is necessary to maintain the heritage value of buildings.

Understanding the mechanisms of collapse of buildings when subjected to seismic activity determines the best method of seismic strengthening and also leads to an understanding of the functions of metallic elements in the reduction of seismic vulnerability.

Even though each element is analysed separately, it is preferable to analyse the behaviour of the whole, seeing that it is the adequate performance of the building as a whole that is sought.

In summary, this study presents several solutions to meet the basic modes of collapse of natural stone masonry structures using metallic elements when subject to seismic activity.

# Index

# Introduction

The present study addresses the most common techniques in using metallic elements for seismic reinforcement. The different types of metallic elements will be examined along with their functions, their properties and installation on a worksite. To better understand these components, we will explain the way masonry buildings collapse in earthquakes, as well as the role played by each of the elements.

This study seeks to highlight the heritage value of buildings and demonstrate less intrusive methods that maintain the architectural value of buildings.

In the Portuguese context these methods have been used for some years, having been brought to Portugal during the French invasions, from 1800 to 1950, approximately. As yet, there are not many documents that discuss methods for seismic reinforcement using metallic elements.

Compared with Portugal, in Italy these elements are widely used in seismic reinforcement. In Italy, because of the frequency of occurrence of earthquakes, there are plenty of studies that demonstrate the capabilities of these elements, much of this study relies on them as well as the retrofitting projects after the 1998 earthquake on the island of Faial in the Azores.

Notwithstanding what has been said, it is possible to obtain a good general knowledge of the use of metallic elements in seismic reinforcement of masonry buildings in the Portugal.

In fact, interest in the retrofitting of heritage buildings is growing both in Portugal and internationally. Accordingly, the issue of seismic safety is a focus of retrofitting. In this context, it is sufficient to recall the earthquake of 1755, which destroyed the city of Lisbon, the Benavente earthquake of 1909, which affected the area of the Tagus Valley, the Azores earthquake of 1980, which partially destroyed the city of Angra do Heroism, and more recently the 1998 earthquake in the Azores, which destroyed much of the islands of Faial and Pico.

Earthquakes have a devastating effect on buildings, as has been demonstrated throughout history and always require major retrofitting work to restore heritage buildings.

It is important to understand the devastating effects are not only derived from the high intensity of earthquakes, but also because the majority of the buildings were not prepared to resist these events, especially masonry buildings.

There are several problems related to assessing the seismic safety of existing buildings, since this is often a rather difficult task because of the difficulty in correctly modelling the structure. Moreover, structural regulations and/or existing reference documents, such as Eurocodes, are prepared for projects for new structures, making it necessary to use adaptations of these texts to make an accurate assessment.

However, in addition to safety assessment and its structural analysis with a view to earthquakes, it is important to conduct a survey of the history of the building, because in addition to ensuring safety, it is necessary to reduce damage to the intrinsic value of the building by using minimally intrusive interventions.

Society attributes cultural value to old buildings, and this value is proportional to age; the older the building, the more cultural value, as a rule. Hence the type of intervention to be carried out in an old building will depend on cultural value attributed to it.

It is necessary that there be a set of guiding principles to successfully retrofit structures, both from the technical and cultural points of view.

The Venice Charter of 1964 is a reference document in the field of patrimonial retrofitting, advocating the adoption of a set of principles, among which are, by their importance, ensuring structural safety, respect for the cultural value of construction, minimal intervention and minimal cost (Santos cit. in Barros et al., 2004).

However, it is not always possible to guarantee the fulfilment of these parameters simultaneously, conflict between these parameters being normal. Sometimes it is impossible to ensure the structural safety without making larger interventions, endangering the cultural value of construction.

It is normal to make use of new techniques and technologies of structural reinforcement in order to find better solutions for the diverse problems that this type of work implies, such as:

- Resistance;

- Transport;

- Implementation on the worksite;

- Easier operation in confined spaces;

- Compatibility between form and function with in existing structures.

In the execution of these projects, it is necessary to assure, in the best way possible, the autonomy of the reinforcing elements from the existing structure through the use of easily removable pre-fabricated elements.

The removability and autonomy of these elements make the maintenance and inspection of the structure easier, preserving the value of both the building in general and its materials.

Figure 1 - Temporary reinforcement of a structure after an earthquake (Source: http://www.dgpatr.pt, June 2005).

The goal of this document is to promote a good understanding of the capabilities of metallic reinforcement elements and the improvements they bring to buildings when subject to seismic events. This understanding will be obtained through the examination of the fundamental design and measurement of metallic elements, the feasibility of implementation and how to correctly mount them.

# 1. Assessment of Seismic Safety and Inspection of the Building

## 1.1. Structural Evaluation

When evaluating the structural performance of a building, one of the first concerns is the acquisition of the data necessary to assessing if the level of safety is acceptable. Therefore, a study should be performed that takes into account the appropriate and recommended parameters for the type of construction being approached.

Structural analysis of old buildings is carried out in the same manner as for new constructions, i.e. by studying the building's behaviour according to a set realistic hypotheses. For this purpose, the effects of actions on different points of the structure are determined and combined, and then are compared to the seismic capacity at those points, taking into account the possibility of the existence of substantial differences.

Regulations and building codes are used in new constructions to reduce uncertainties, both in the level of seismic capacity and with respect to actions. The safety coefficients used aim to increase the safety of the structure, without disproportionately increasing the dimensions of structural elements and their costs. In contrast, uncertainty is greater in retrofitting, both in terms of efficiency as in the final cost.

In terms of modelling the behaviour of older structures it is also more difficult and conditioned by various factors such as:

- The difficulty of performing a proper survey of the structure;

- Uncertainties regarding the characteristics of existing materials;

- Not knowing of previous alterations or repairs (the events of the past not always being visible in the structure of the present moment).

## 1.2. Safety Level of Buildings

As previously mentioned, there are differences between the evaluation of new buildings and old buildings. Therefore, the assessment of the safety of old buildings takes into account their cultural value, so it is necessary to evaluate the costs and benefits associated with the retrofitting project.

The benefit is a result of risk reduction, while the cost is associated with the value of the retrofitting project, as well as the variation of its cultural value.

Given the apparent increase in uncertainty in the forces and resistance of materials, the partial safety coefficients to be taken into account should normally be higher than those used in a new construction, because of the need to err on the side of safety. It is evident that if an exhaustive survey of permanent loads and resistance of materials is made, then we can adopt safety factors lower than those for new construction if the level of confidence is clearly high.

## 1.3. Building Inspection

In all retrofitting projects it is necessary to inspect the building so that sufficient data for a correct analysis can be obtained. Good execution of retrofitting projects is only possible with this information.

Inspections related to structural retrofitting of old buildings will carried out in two phases:

- A preliminary inspection;

- A detailed inspection following the first.

Since the preliminary inspection is qualitative, it uses visual observation or simple equipment in the determination of existing problems. On the other hand, the detailed inspection is quantitative in nature; tests and measurements are performed to determine the characteristics of the materials making up the structure and the dynamic properties of the structure as a whole. This last inspection should also serve to classify the types of the most important structural anomalies.

Figure 2 – Survey of rebar with rebar detector.

In the majority of cases, in these inspections it is equally necessary to evaluate the layout of the building because there are not normally plans for old buildings.

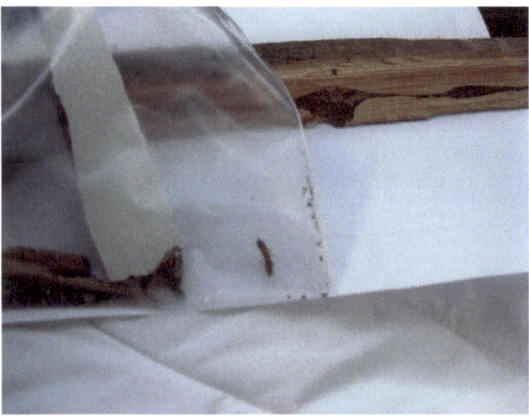

Figure 3 – Laboratory testing for the presence of termites in a wooden beam (Source: M. M. T. E.C.).

When the level of detail of the inspection is high, it is necessary to use laboratory testing to gain a deeper understanding of the materials and structure.

## 2. Modelling the Structure of the Building

### 2.1. Generalities

The modelling of the behaviour of the structures of old buildings employs the usual methods of structural analysis, where, assuming certain hypotheses (linear elastic response, the use of nonlinear models, etc..), information is sought on the existing states of tension, or those which could be produced in the various parts of the structure (Santos cit. in Barros et al., 2004).

It should be noted, however, that, at the risk of simplification, given the possible lack of knowledge regarding the actual characteristics of the materials it is generally necessary to note in the representation of the operation of older structures that in principle the results will always less reliable than in the case of new structures (Santos cit. in Barros et al., 2004).

The behaviour of any structure is influenced by three main factors:

- The shape and joints of the structure;

- The materials of construction;

- Forces, accelerations and deformations imposed (actions) (ICOMOS, 2004).

### 2.2. Structural Plan and Damage

A structural typology represents the behaviour of diverse structures to various actions existing in the building, making possible the determination of the form of structure that will guarantee its stability.

A deep knowledge of the behaviour of the structure is necessary to precisely delineate this schema; it is necessary to know the following:

- The materials;

- The size of the elements;

- The condition of the ground;

- The joints between different elements;

- The anomalies, etc.

In the specific case of masonry structures is necessary to be aware that these are usually made of materials having a very low tensile strength and can easily exhibit internal cracking or separation between elements. However, these signs are not necessarily an indication of danger because masonry structures mainly work through compression (ICOMOS, 2004).

In general, masonry structures depend on the effect of floors or roofs to distribute the lateral loads and thus behaviour ensure the stability of the structure. It is also necessary to understand the sequence of construction, because the different characteristics of different periods of masonry may affect the overall behaviour of the structure (ICOMOS, 2004).

Thick walls and especially the double walls should be noted. This stems from the fact that their cores are sometimes of poor quality; various problems arise such as:

- Vertical cracks;

- Strains and detachments of outer surface (which can lead to the collapse of the structure).

Therefore the behaviour of the structure and associated phenomena must be adequately represented to make the use of computational tools possible.

Although only simplified models can be used in certain situations, for example, in simple terms of static equilibrium the contemporary structural analysis of old buildings, especially when subjected to the action seismic events, is based on sophisticated models. The difficulty of manual calculation is understood, and automatic calculation programs are almost always used, knitting together of finite elements in the representation of the behaviour of diverse structural elements.

The preparation of the structural typologies should be based on the survey of the construction and its surroundings as indicated above. If available, already existing information on the building can also be used (memories, drawings, photos, etc...), although at least partial confirmation of this data should be obtained. As a complement, it is also important to have contact with the neighbouring population in order to obtain a better understanding of the history and events associated with the building.

The scheme should consider any changes and degradations suffered over time, the effects of which may influence the overall behaviour of the structure by altering the distribution of forces. These changes can be caused by both natural phenomena and by human activity.

In the modelling of complex structures, a single structural schema should not be used, but rather several alternative or complementary structural schema. It must be remembered, though, that different structural schemes could lead to results with substantially different values, even concerning the same structural element (Santos cit. In Barros et al., 2004).

If in-situ tests determining the dynamic properties of the structure (natural frequencies, damping, etc...) have been performed, the results obtained can be compared to the the values attained through modelling, which will aid in refining the model of the structure.

14

## 2.3. Characteristics of Materials

The properties of the materials (particularly resistance), which are basic parameters for any calculations, can be degraded by the action of chemical, physical or biological processes. The rate of degradation depends on material properties (such as porosity) and the existing protection (protruding roof, etc.) and (lack of) maintenance. Although the degradation can manifest itself on the surface and be immediately visible via a surface inspection (blooming, high porosity, etc.), there are also degradation processes that can only be detected by more sophisticated means (termite attack on wood, etc...) (ICOMOS, 2004).

As mentioned in the previous paragraph, tests on structural materials bring to light their characteristics and these characteristics can also be obtained from existing data bases though one should be on guard against differences that may occur in any particular case.

The results obtained should permit the quantification of characteristic values of the properties of structural materials. Assuming normal distributions permits characteristic values be derived from the average value, taking into account the coefficient of variation and the sample size. However, as the sampling is generally not very extensive, sometimes a criterion adopted is derive the characteristic value from the minimum value of the sample. There should, in any case, be at least two values (Santos cit. In Barros et al, 2004).

As for the partial safety coefficients of materials, these should be related to the uncertainty associated with the determination of characteristic values of the resistances. Although dependent on the quality of information used, since the actual values are known, lower values than those generally envisaged can generally be adopted in the design of new structures, values of around 1.1 for steel being recommended, from 1.2 to 1.3 for concrete and wood, and from 1.3 to 1.5 for masonry (Santos cit. in Barros et al., 2004).

15

## 2.4. Actions on the Structure

The permanent loads (weights themselves, etc.) should in principle be obtained from the survey geometry and the constitution of the construction. Items of existing information (drawings, etc.) may also be useful but should be used carefully (Santos cit. In Barros et al, 2004).

Actions are defined as any agent (forces, deformations, etc.) that produce stress and strain in the structure and any phenomenon (chemical, biological, etc.) affecting materials, usually reducing its strength. The original actions, occurring since the start of construction until its completion (eg, weight), may be modified during their lifetime and often these changes produce wear and tear (ICOMOS, 2004).

Before making a decision about any repairs to be made to the structure, it is always necessary to have a perfect knowledge of the application or modification of the original actions, and these can be diverse in nature and, thus cause different changes both in the structure and materials.

In this study dynamic mechanical actions will be treated more specifically because dynamic action is more important and is normally caused by seismic events.

The mechanical actions acting on the structure produce stresses and strains in the material, possibly resulting in cracking, crushing and visible movement. Dynamic actions are produced when a structure is subjected to accelerations resulting from seismic events, wind, hurricanes, mechanical vibration, etc. (ICOMOS, 2004).

In the dynamic action resulting from an earthquake, the intensity of the forces produced is related to both the magnitude of acceleration and the natural frequencies of the structure and its ability to dissipate energy. The effect of an earthquake is also related to the previous history of earthquakes, which may have progressively weakened the structure (ICOMOS, 2004).

Thus, the action of earthquakes is in principle measured in the same manner as in the design of new buildings, and in certain cases can be admitted with a shorter reference time period than in the adoption of new structures, permitting the reduction of the value of the action. In certain situations it may also be possible to determine a localised seismic zoning, which eventually will reduce the value of this action (Santos cit. In Barros et al., 2004).

The coefficients of the behaviour to adopt in the case of linear seismic analysis should be established taking default values recommended in the draft regulations for new structures, taking into account the specific layout of the old structure. In the case of masonry buildings, for example, depending on the mechanical characteristics of the blocks and mortar used, and the possible presence of metallic connecting elements, values around 1.5-2.0 are recommended (Santos cit. In Barros et al., 2004).

 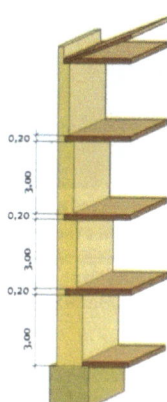

Figure 4 – Building with Pombaline metalwork, in this case a higher value can be adopted in the order of 2.0-2.5, depending on the proportion of metalworked walls conjoining the walls of the building (Appleton, 2003).

Regarding the partial safety coefficients of the actions, as well as the coefficient of performance, the regulatory values pertaining to the design of new constructions should be adopted or, preferentially, conservative values, erring on the side of safety.

The same is true in the case of permanent loads, in which it is also favourable to use values considered superior to those used for new structures. Notwithstanding what has been said above, and particularly if the values being considered have been obtained from an

17

exhaustive assessment of the building, coefficients in the order of 1.2 can be considered. (Santos cit. In Barros et al., 2004).

## 3. Determining the Type of Intervention

After a phase of survey and diagnosis of the building's problems it is necessary to determine the best intervention in order to structurally retrofit the building to enable it to resist seismic action.

The safety evaluation of an old building structure is carried out as is done for the new structures, comparing the calculated values of the resistances of the materials with calculated values of the effects of actions in every point of the structure (Santos cit . in Barros et al., 2004).

As noted above, the results of modelling the structures of old buildings are not, however, and generally as reliable as in the case of new buildings, which should be evaluated by the consistency of these results with the actual state of the building , particularly concerning the possible existence of damage (Santos cit. in Barros et al., 2004).

Therefore, it is extremely important to consider all possible methods of evaluation so that one can make decisions free of errors.

In the phase of assessment, as already pointed out, it is important to gather all existing information on the history of the building, changes throughout its life course, anomalies, the performance during past earthquakes, knowledge of the surrounding buildings, etc....

Information about the building's history allows one to evaluate the results of the performance of the structure during seismic events and thereby help predict its future behaviour. Indeed, in old masonry buildings is often difficult to predict the functioning of the joints between the floors, usually in wood, and masonry walls.

Changes and extensions made by the owners are also a problem due to the difficulty of predicting its performance. Hence knowledge of the results of similar phenomena found in the present project will facilitate decision making.

Sometimes, companies in the industry help work along by comparing the building in question with others like it for which a study has been carried out.

So with all these analyses (historical, qualitative and quantitative) a good evaluation of a building's seismic vulnerability can be prepared so as to be able to determine the best retrofitting strategy.

If it is not possible to draw clear conclusions regarding an eventual lack of safety in the structure or of some of its elements, it is always preferable not to intervene in order to maintain the highest level of cultural value for the building. Furthermore, the cost of intervention will be reduced if its retrofit or structural reinforcement were not done (Santos cit. in Barros et al., 2004).

After all the analysis, and after having determined an number of different strategies of rehabilitation, a report should be prepared which addresses all the details that were considered in the analysis. With this systematic evaluation of parameters, the options of intervention will be properly justified.

Figure 5 shows a simplified flowchart of actions in developing the structural evaluation of old masonry buildings.

Figure 5 - Flowchart of our actions in the structural evaluation of old buildings (Adapted from Santos, 2004).

# 4. Components of Seismic Retrofitting

## 4.1. Assumptions of the Reinforcement Project

Correct organisation of the structural mesh is the first and most important requirement of any masonry building, constructed according to the state of the art. A good structural mesh will have always two lines of orthogonally placed walls which delimit rectangular environments whose dimensions are determined in relation to wall thickness. On the other hand, structural dispositions which have large gaps or, even worse, are not closed on all four sides, are situations that are intrinsically weaker which will inevitably experience crisis in the occurrence of a seismic event.

Figure 6 - The "Anti-seismic Home" Logorio of Pyrrhus, The Security of Buildings Against Earthquakes, 1570 (Source. M. M. T. E.C.).

The awareness of the fundamental role played by the structural design of a building is present in "Trattati classici de arquitecttura", by Vitruvius and L. B. Alberti a Palladio, explicitly and consequently assuming the first anti-seismic house, proposed by Pirro Ligorio after the earthquake of Ferrara in 1570 (Figure 6).

The same awareness is behind the famous drawing of Milízia (1781), which speaks of an anti-seismic "home" as a "box" of timber may only permit changes to the whole, but not interior changes that may compromise the integrity of the structure.

In the spirit of the classical texts, modern seismic standards take up the requirement to limit the maximum gap in a free wall facade admissible in a masonry building.

## 4.2. Retrofitting Projects

The retrofitting project of a building is a whole similar to the project of a new work and must contain the documents necessary to its proper execution, both drawings and descriptions.

However, retrofitting projects, due to the degree of their specialisation, should be developed with great detail regarding the method of execution of the work and with an eye to effectiveness at every point. In this way, the possibility of errors will be reduced. One should also mention the equipment and materials used and their conditions of application.

In the case of buildings that are still used during the execution of the work, such a situation should also be taken into account. A realistic estimation of the cost of the job should also be included.

The solutions of repair or reinforcement of old buildings are very varied, depending on the type of problem that one seeks to correct, particularly if the repair degraded materials is necessary, or the repair the effects of mechanical actions, or even work to improve safety in the face of seismic events (Santos cit. in Barros et al., 2004).

The choice of the solutions should therefore be justified and be subject to cost-benefit analyses. Clearly maximum efficiency at the lowest cost should be kept in mind, respecting as much as possible the principles already mentioned, particularly respect for the cultural value of the building (Santos cit. in Barros et al., 2004).

Sometimes more intrusive methods, such as the application of "cordolo", a metal component that will be subject to a thorough analysis further ahead, do not fit well into a strict cost-benefit analysis. In the case of the retrofitting of houses affected by the earthquake in the Azores in 1998, the application of "cordolo" was abandoned because its implementation was of a very high degree of difficulty and the costs did not fit in well to a cost-benefit analysis.

## 4.3. The Use of Steel

Today, even though there are various materials in use for reinforcing structures, steel is the most used material in structural reinforcement where masonry structures subject to the action of seismic events are concerned.

Its main feature is its high ductility, a property of maintaining resistance under significant deformation, the reason for it being suggested for use in structures located in seismic zones. In fact, reenforcing elements made of this material, make likely a more gradual collapse in the event of a structural failure.

Steel is a material with great flexibility of use in construction, which permits the successful resolution of structural problems given the various forms in which it comes to market, be they geometrical (bars, rolled, cold formed, tubular sections, sheet metal) or mechanical (various types of steel and resistance classes) (http://www.dgpatr.pt, June 2005).

The possibilities for this material are so large that they allow the implementation of a wide range of operations, ranging from simple reinforcement using one element up to the complete restructuring and global anti-seismic modification of the structure (http://dgpatr.pt, June 2005).

23

The use of reinforcing steel elements in the structure has various advantages, such as:

1. Aesthetic aspects, e. g. the slenderness and clarity of forms;

2. Possibility of modelling;

3. Reversibility.

It also has several advantages for use on site, such as:

1. Its small size;

2. Its "lightness" when considering the binomial weight-resistance and bearing its own weight;

3. Simplicity of transport;

4. Use in confined spaces;

5. Ease of installation on site;

6. Reduced turnarounds;

7. Prefabrication.

In the latter respect, prefabrication, its use reduces the time of assembly of the elements, since they are implemented in a workshop, and enables a higher rate of parallel production in processing the work, as well as better quality control. The structural elements arrive at the worksite in parts which are easily mounted using screwed joints. In the following pictures you can see put into work the components of a prefabricated truss.

Reversibility allows the reuse of metallic elements, which is a great advantage in various types of project, such as use as a temporary reinforcement in buildings that have suffered an earthquake.

The lightness of metallic elements facilitates the placing and mounting work, it being possible to transport and mount the elements manually, as seen in Figure 7.

Figure 7 - Components of a truss being manually unloaded at a worksite (Source:M. M. T. E. C.).

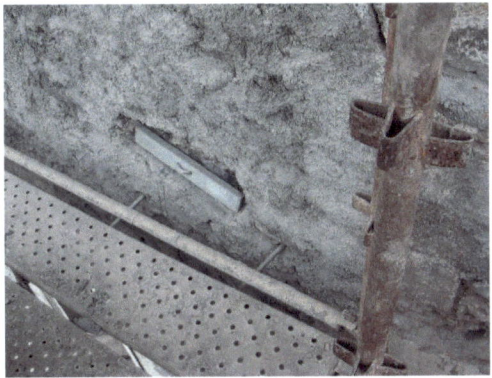

Figure 8 - Rod anchoring bar, a lightweight and less intrusive component (Source: M. M. T. E. C.).

The reduced size of these elements, along with their elevated resistance, allow the execution of retrofitting projects in buildings with cultural value without using more intrusive methods. Good performance of the building structure is thus obtained without affecting its cultural and heritage value.

The ease and speed of placement in the steel work makes it an ideal instrument for strengthening structures, especially when it acts in buildings that have suffered an earthquake and are vulnerable. Rapid action can sometimes prevent the total collapse of the building.

## 4.4. Levels of Interaction

As we have seen earlier, there are several factors that can influence the type of intervention to be applied. Thus, there are several possible levels of intervention, and these vary with the factors previously mentioned, such as the historical value, the existing conditions, the rate of seismic vulnerability, available funds, etc..

The levels of intervention can be classified as:

· Safety;

· Repair;

· Reinforcement;

· Restructuring.

The first two levels presuppose the existence of improper functioning of the structure, from the standpoint of safety; the latter two, while not necessarily implying the existence of structural damage that jeopardises safety, are appropriate when the intention is to give a new purpose to the structure, or to modify it in order to comply with new regulations (http://www.dgpatr.pt, June 2005).

### 4.4.1. Safety

By safeguarding, is meant a set of interventions, usually temporary in nature, aimed at ensuring the safety of the structure until the definitive retrofit of the structure is completed. This type of intervention is generally used where there is a possibility of partial or total collapse of the structure (http://www.dgpatr.pt, June 2005).

A typical case for resort to safeguards is when emergency measures are taken in the aftermath of an earthquake. It applies, therefore, when the urgency of intervention is a priority, and the lack of materials and funds for financing measures requires simple and rapid action executed with great operational flexibility (http://www.dgpatr.pt, June 2005.).

### 4.4.2. Repair

Repair takes place after the safeguarding and involves the execution of work in order to restore security and initial functionality to the structure. This type of work is carried out after the functional abnormalities caused by, for example, the effects of atmospheric or seismic events, or other factors causing structural damage and compromise the safety of buildings (http://www.dgpatr.pt, June 2005).

In contrast to safeguarding, repair has a permanent character, given that it is sometimes easy to predict when the structural damage is due to aging effects of the structure or the effects of a more punctual nature. In this case, the diagnosis is usually easy, requiring urgent intervention measures (http://www.dgpatr.pt, June 2005).

### 4.4.3. Reinforcement

The next level of intervention is classified as reinforcement. Here it is not necessary to have structural damage, but rather the need to give the structure sufficient strength to cope with current or new uses. For example, higher loads or the need to provide seismic resistance as in the case of buildings built in the past when there were no requirements of this nature (http://www.dgpatr.pt, June 2005).

Figure 9 – Detail of the support device used to prevent the transmission of impulses to the walls (Source: http://www.dgpatr.pt, June 2005).

Figure 10 – Reinforcement including bracing to improve seismic resistance (Source: http://www.dgpatr.pt, June 2005).

In general, reinforcement does not provide significant changes in the structural design of resistance. Similar to repair, structural reinforcement can have varying levels of intensity conforming to the level of resistance required of the building. This is particularly important from the seismic point of view when it is important to improve or properly adapt the structure with this in mind.

### 4.4.4. Restructuring

The most general and complex intervention is restructuring. It consists in modifying all or part of the space, the volume and the design of structural resistance. This type of intervention is performed when you want a new distribution of spaces, or, when faced with new regulations, the existing structural design is inadequate, though strengthened, to the new function of the structure (http://www.dgpatr.pt, June 2005).

The following work can be considered part of restructuring (http://www.dgpatr.pt, June 2005).

1 Evacuation of the interior structure of the initial construction, with subsequent insertion of a new structure inside;

2 Expansions, either horizontal or vertical;

3 Relaxation of the structure and insertion of new sub-structures within the existing structure.

In seismic zones, such as Portugal, all these types of restructuring require that new structures have adequate resistance to seismic action (http://www.dgpatr.pt, June 2005).

## 4.5. Economic considerations

Retrofitting methods are increasingly being used in Portugal today since these methods are cheaper than engaging in new construction.

When contemplating the structural retrofit of a building it is necessary to prepare economic studies of the project to establish economic viability and the methods to apply. However, the budgeting of such works is a problem for practitioners, mainly due to lack of definition of costs to be applied to operations.

Publications with the values of the costs of retrofitting work do not yet exist in Portugal. The National Laboratory for Civil Engineering (LNEC) has started preparing hundreds of sheets of costs of this type of work which should appear shortly. Meanwhile practitioners need to prepare estimations and budgets without aids case by case, using tables they have developed themselves or by using other types of information such as: analytical sheets of reference cost composition or sheets of average prices (ICOMOS, 2004).

Reinforcement with metallic structures is one of the most used methods in Portugal, beginning to be used between 1800 and 1950 and having a wide usage today all over the country.

According to the Heritage Department, the structural retrofitting with metallic elements, taking into account the vast italian experience in this area, can be separated out into individual operations on:

1. Walls;

2. Cantilevered balconies and terraces;

3. Floors;

4. Roofs;

5. Stairs;

6. Structures inserted in the interior;

7. Other types of intervention.

It is interesting to note that the roofs have the highest percentage cost and are most frequent among intervention operations. This will be due to the high use of metallic elements in the connections between the various constituent parts of the roofs.

Floors come in second place, due to the fact that old buildings normally have wooden floors which are reinforced by steel beams, sheets of shaped metal, etc.

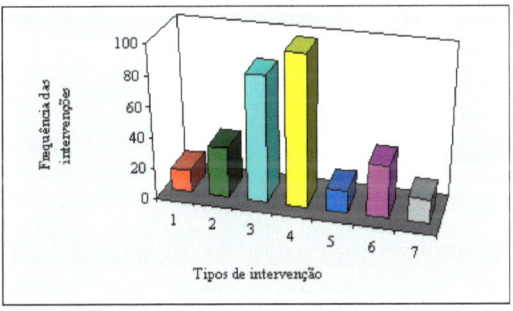

Figure 11 - Frequency distribution by type of intervention (Source: http://www.dgpatr.pt, June 2005).

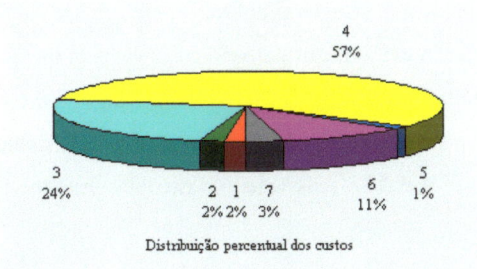

Figure 12 - Costs by type as a percentage of interventions (Source: http://www.dgpatr.pt, June 2005).

In other portions the use of metallic elements has increased because, when one is speaking of seismic reinforcement with metallic elements, the best performance of a building is obtained from all the parts of the building working together. Note that this working together is only possible to achieve when the reinforcement is made at all levels of the building.

It is also important to mention that the increase in the use of metallic elements is due to fact that they make less intrusive and less expensive interventions possible.

## 5. Typical Types of Collapse for Masonry Structures

It is important to keep in mind that the type of building collapse can vary according to circumstances and that because of this reinforcements used must vary as well. This being the case, two basic types of collapse for stone masonry structures will be defined in what follows.

## 5.1. First Type of Collapse Mechanism

These types of collapse pertain to the main facade, but can easily be extended to the rear facade as well (whether of not this facade has different dimensions). The safety analysis comes from a detailed analysis of the stability of the main facade.

The first two mechanisms are grossly simplified: both assume an ideal wall with no openings and only differ in the joints of perpendicular walls of the gables. If there is no connection, the wall falls as one whole, if there is a connection the facade wall will be damaged around the connection; this is the classic failure mechanism known as a Rondelet. Rondelet described this in his Tract of Principles from the 19th century.

The third mechanism is a modification of the Rondelet mechanism, which takes into account the presence of openings for doors and windows in the facade.

In the case of a facade wall, the first mechanism has an acceleration, which following the R. S. A. is:

$$a = \beta \times g = (S/H) \times g$$

Where:

1   $a$ – acceleration;

2   $g$ – acceleration of gravity;

3   $S$ – thickness of the facade;

4    H – height of the building.

While the second mechanism requires, in the case where the entire wall is involved, only the last three levels excluding the buried level where the accelerations are the same, respectively, to:

$$a = \beta \times g = 1.5 \times (S/H) \times g$$

In the Rondelet mechanism, the twisting of each half of the wall comes from the point of inclined fissure, which forms an angle $\theta$ with the vertical.  As a consequence, the force that opposes the twisting motion is the component (P·sin$\theta$), perpendicular to the line of the fissure. This corresponds to the weight of the triangular portion of the effected wall, while the force that moves the mechanism is the seismic force applied to that same triangular portion.

The weight is the mechanism of global twisting (S/2) while the seismic force is equal to (H/3)·sin$\theta$ (Figure 13).

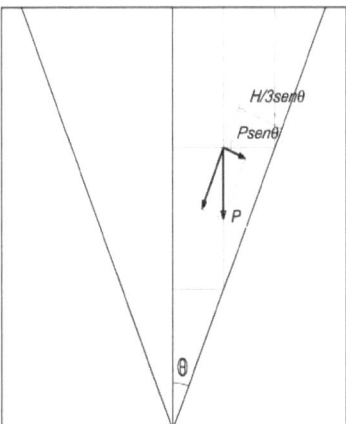

Figure 13 – Rondelet first type mechanism (Source: M. M. T. E. C.).

It is obvious that the proposed calculation for the second type mechanism ignores the removal of the "fabric" of the wall of the building necessary to reestablish, along the lines of the fracture, the geometric congruence of the mechanism, in this way obtaining the multiplier values of the collapse which could be inferior to the real values.

However it must be remembered that even in the calculation of the first type mechanism neglects the stabilising effect due to the connection with the walls or any bracing. This means that the resistance associated with such a mechanism is, in fact, higher than that obtained by the above calculation, or: not considering these factors is an approach which will eventually be seen as defensive, but safe.

On the other hand, real walls always have openings in the facade which modify the Rondelet mechanism. The effect of such openings involves a reduction of the stabilising work which follows from the rocks of the facade falling and, as a consequence, an additional reduction of the difference between multipliers of collapse of the first and second type mechanisms.

In conclusion, measuring the seismic resistance of facade walls where the whole wall falls, assuming no openings, is a safe measure but not excessively cautious (taking into account seismic acceleration capable of damaging the building and, potentially, its collapse).

The drop caused by the collapse of a vulnerable facade is caused by the activated mechanism during a seismic event.

The installation of metal moorings at the level of the floors, rods ensure the connection of the facades to bracing walls (side walls of the building), allowing the transfer of the seismic action to the latter. Moreover, as the bracing walls oppose the action of falling produced by the earthquake, their greater size can provide a magnitude of resistance above that of the walls of the facade.

With a well-developed structural mesh of walls, the traditional approach to reinforcement, tightly securing the facades using rods, creates two known problems.

First, it is clear that it is not sufficient to put the rods in line with the bracing walls, which is where the rods will be suitably anchored. This disposition of rods only impedes the collapse of the wall as a single unit, but doesn't prevent the Rondelet mechanisms of partial collapse from operating.

In fact, bending is induced in the segment of wall between two rods by seismic action. However, because of the very modest tensile strength of the stone masonry, the only option of the wall for opposing this stress is to bow along this segment of wall and the thickness of the wall.

Too large a distance between the rods may create such a bowing, and this itself can be responsible for failure by flexion in the wall (Rondelet mechanism of collapse).

The optimal spacing between the rods depends, naturally, on the thickness and amount of wall to secure and not only the disposition of door and window openings.

The thickness and quality of wall directly define the nature of the bowing while the disposition of the openings conditions the possibility of effective anchoring of the rods.

In cases where intermediate rods are needed adjacent to the bracing walls, a problem arises of transferring the action of containment of the facade by the rods to the bracing walls.

Different solutions are naturally possible, and they depend more on the gaps in the facade to be sustained.

Modest gaps can be effectively dealt with using inclined rods while for larger gaps the solution has to be more developed, through the insertion of a rectangular beam where intermediate rods will be fastened.

## 5.2. Second Type of Collapse Mechanism

Seismic action which is transferred through connected rods from the facade to the reticulated beam should be supported by the bracing walls where the beam rests.

In this way, as was said, the mechanical resistance of the bracing walls is considerably greater than that of the facade walls and if it is exceeded is the origin of the classical diagonal rupture, the second type of mechanism of collapse (Figure 14).

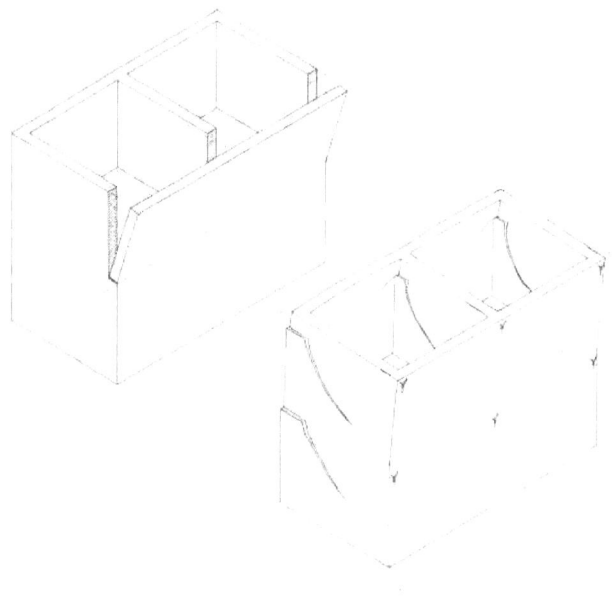

Figure 14 – First type of collapse mechanism (non-connected facade) and the second type (facade secured by rods) (Source: M. M. T. E. C.).

The formal verification of mechanisms of the second type or the calculation of bracing walls' seismic resistance is a problem for which there remains great uncertainty today and, it is not by chance that the role of subjective assumptions continues to be relevant in all available models.

The procedures for seismic action based on the calculation of the tangential requirements on parallel walls suggests the definition of a parameter of resistance to be considered which in the absence of determinations based on experimental results assumes a strongly conventional nature.

Also, the procedures based on the collapse mechanisms are heavily conditioned by the choice of the most probable mechanism, only precise experimentation can guarantee the confidence of verification.

Nevertheless the comparison of verifications which involve the evaluation of mechanical resistance, experimentation based in mechanisms of collapse has the indisputable merit of modelling the real structural behaviour of buildings with greater fidelity, problems of stability being most urgent among the problems of mechanical resistance.

In order to rationally define the mechanisms of the second type it can be useful to rely on laboratory experimentation.

Figure 15 compares the second type of mechanisms of collapse of a single wall, without openings but with securing rods (installed on two levels).

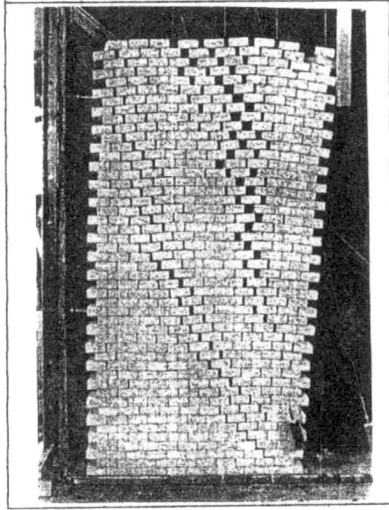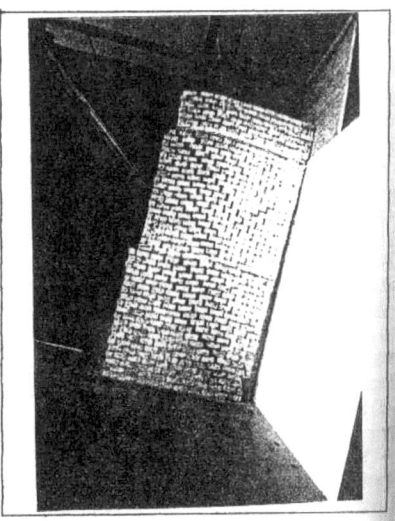

Figure 15 – The effect of rods on collapse mechanisms of the second type: collapse mechanisms without (left) and with (right) rods (Carocci et al 2000).

37

# 6. Metallic Seismic Reinforcement Elements

Throughout this chapter the different metallic elements used in the seismic reinforcement of masonry buildings will be presented. A case study will be presented in Chapter 7 to provide a more thorough understanding of their functions and real-life application; it is important to analyse the combination of masonry with metallic elements.

Technical drawings of each type of component and photos of worksite application will be included.

## 6.1. Rods

The rod is a component of structural reinforcement steel, usually stainless steel, which works through tension and may have active or passive roles in the structure. However, its function, in the case of anti-seismic reinforcement, is generally passive, only providing tensile action in the event of a seismic event.

Figure 16 – Rods with turnbuckles in a line, Horta Hospital, Faial Island, Azores (Source: M. M. T. E. C.).

This member has been applied now for many years in Portugal. In times past, they were only applied in buildings where the owners were experiencing economic pressure or in buildings where an elevated level of seismic capacity was necessary consonant with the building's function, as in the case of hospitals (Figure 16).

There are various types of rods, taking the form of screws, cables, bars or rebar of iron or steel. For seismic activity, iron rods in the form of screws were used in the past where today steel cables similar to pre-stressing cables are normally used.

Figure 17 – On the left, rods in the form of screws and on the right rods in the form of cables (Source: M. M. T. E. C.).

The application of rods creates resistance to horizontal seismic forces, perpendicular to the facade walls, preventing their collapse through the tension created by the rods.

In the reinforcement of small buildings these rods are applied throughout the space and are anchored in the facades. In buildings with an elevated number of gaps, it is necessary to have an intermediate anchorage, implemented through a trussed beam.

After anchoring the cable in the facades, the tension is regulated using a turnbuckle (Figure 17). Such a device allows the increase and reduction of cable tension and is generally manually adjustable. However, the cable should not exert any tension on the facades.

Figure 18 – On the left, a rod with intermediate connection to a trussed beam, on the right, a rod without intermediate connection on the exterior part of the facade (Source: M. M. T. E. C.).

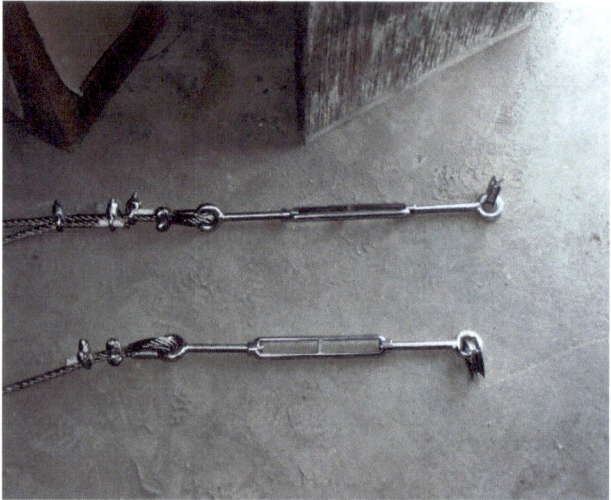

Figure 19 – Turnbuckles (Source: M. M. T. E. C.).

As noted, the rods are passive, not exercising tension on the facades, and only become active during a seismic event. When a seismic event occurs the rods permit all the walls of the facade, through the connection, to work together.

It is important to note the possibility of regulating the tension of the rods because the occurrence of a new earthquake will require resetting the tension of the rods to enable resistance to a following earthquake.

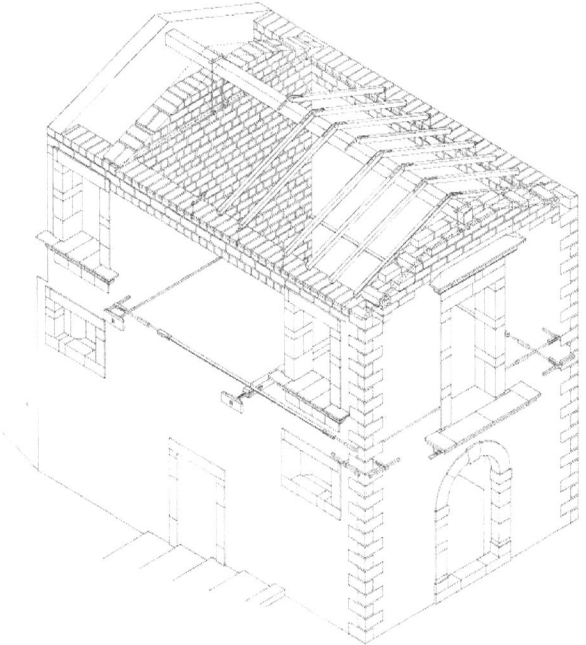

Figure 20 – Application design for rods in a building in Ortigia, Italy (Carocci et al 2000).

### 6.2. Crowning Belt ("cordolo")

The use of a "cordolo", better known as a crowning belt, is not very common in Portugal. Nevertheless, it is a common practice in the seismic reinforcement of masonry buildings in Italy.

Recently this method was applied to some houses in the retrofit following the earthquake of 1998 on the island of Faial in the Azores. Although it was not possible to do this in all buildings because of the method's high level of difficulty. Not only is it necessary to have skilled labour, it requires some a longer time to complete the job, which makes the project more expensive.

The purpose of this method is to divide the horizontal seismic forces connecting the walls in such a way as to make them behave as a box, distributing vertical loads and

41

reducing the movement of the roof, i.e. confining and securing the whole structure of masonry and wood.

The "cordolo", essentially, is a metallic bar which is applied to the top of buildings, linking the four facades at the same time as linking them to the roof. The goal is to have this set of structures function simultaneously as a unit when an earthquake occurs.

Figure 21 - Preparing the top of the walls for the installation of the "cordolo" (Source: M. M. T. E. C.).

This method comprises a reinforcement bar of φ 24mm for walls 45cm to 50cm in thickness, the diameter varying with thickness of the wall. This bar must be inserted into the interior of the walls at least 50cm from the top, should pass through all the walls, making the connection between them like a crown.

Figure 22 – Installation of a "cordolo" in progress (Source: M. M. T. E. C.).

The connection between the bar and the roof is done through rods in the form of hooks that link the male end of the hook and its opposite by screws that will fix metal plates to the legs of the roof.

Figure 23 - Connecting the "cordolo" to the roof (Source: M. M. T. E. C.).

The same reinforcement should be done on the side walls, if any, to prevent the collapse of the gable peak, which has a higher degree of fragility.

As can be seen, the implementation of this work using this method is complex and intrusive which imbalances cost/benefit ratios.

## 6.3. Anchoring Bars

Anchoring bars are elements, usually of steel, which anchor the rods on the exterior part of the walls. These bars can be rectangular or round and about 50cm in diameter, respectively.

They are subjected to a linear force, obtained after the distribution of tensile forces transmitted through the rods to the walls.

The anchors can be implemented in two ways, depending on the situation and the type of building.

If there is no chance of keeping the anchoring bars out of sight for esthetic reasons, then there are two ways of implementing the member without being too intrusive.

When possible it is possible to cement the entire facade at the end of the project, then a gap-shaped mold with the dimensions of the anchoring bar will be made in the place where the rods will be moored. This method, while it doesn't leave the tie bar visible, does not permit access to the anchoring bar over time which will make any subsequent work on the bar intrusive.

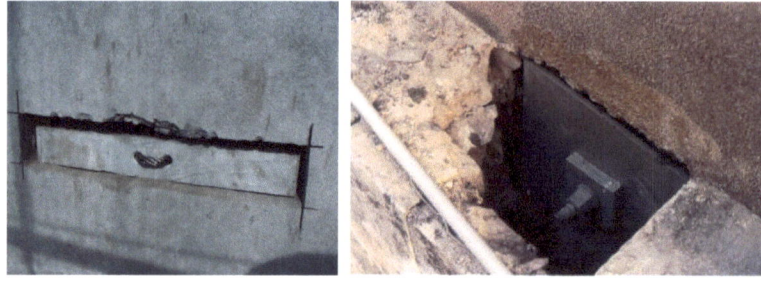

Figure 24 - Anchoring bars (Source: M. M. T. E. C.).

Figure 25 - Anchoring Bars (Source: M. M. T. E. C.).

Another possibility occurs when the facades are visible stones, which sometimes allows you to disguise the anchoring bar between the stones of the facade. This is simpler and more easily performed the above method.

While it is possible to keep the bars in sight, the bars then need to anchor the rods on the exterior face of the buildings, this method requires the building to be cemented later.

## 6.4. Trussed Beam (Trellis)

This member is composed of metallic parts in hollow rectangular cross-sections and is normally pre-fabricated. This saves time in the execution of the project and facilitates mounting. The beam is unloaded at the worksite in pieces which permits the control of its size at the time of mounting, adapting to the gap in the building.

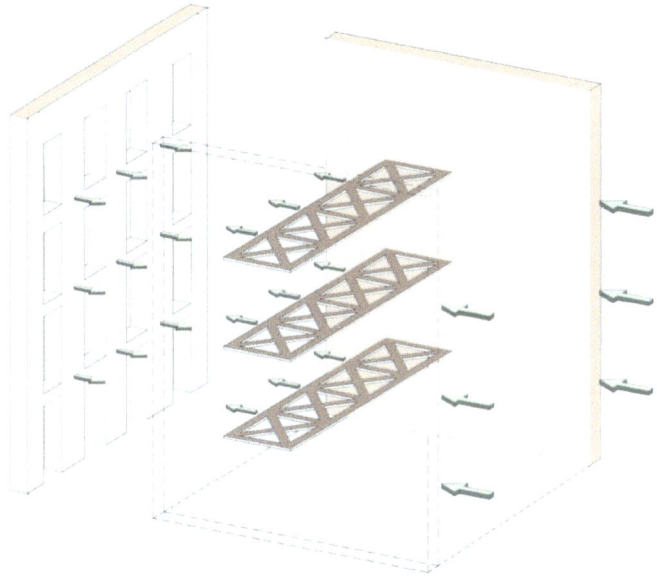

Figure 26 – A trussed beam controlling the facade, transferring the seismic forces applied to the facade to the bracing walls through the rods (Source: M. M. T. E. C.).

The trussed beam receives, preferably but not necessarily, along with those elements that are perpendicular to the truss, the forces transmitted by the intermediary rods and transfers them to its ends, the bracing walls where it is anchored (Figure 26).

In cases where it is possible to place a trussed beam in a central position with respect to the bracing walls (the side walls of the building), the transmission of loads will happen in the same way whether the forces come from either direction, from the front or from the rear facade. The alignment of the trussed beam with both the facades should be taken into account.

Figure 27 – trussed beam placed in a central position relative to the side walls of the building (Source: M. M. T. E. C.).

In cases where one is forced to place a trussed beam in a non-central position, it is preferable that it be placed as far as possible from the facade wall to be secured. This will allow more of the bracing wall and its greater weight to be mobilised. This provision will significantly increase the resistance to collapse brought on by the action of the intermediary rods which are connected to the trussed beam.

The specifications of the trussed beam are normally calculated by computer programs.

Figure 28 – Left, connection of a trussed beam to the side walls; right, connection of the rod to the trussed beam (Source: M. M. T. E. C.).

## 6.5. Screwed Joints

The screwed joints, as its name implies, are joints made by metal screws which connect the metal elements with the existing building structure.

These joints are extremely important because they ensure the safety and smooth operation of the other metallic elements used in the seismic reinforcement of the building.

In order to guarantee the good performance of these screwed joints it is necessary to correctly calculate the sizing of the links of rods to metal faces and connecting the lattice of metal faces to the beams of wooden floors, among other links.

Figure 29 - Different types of screwed joints (Source: M. M. T. E. C.).

In implementing screwed joints the number of screws necessary is determined through the sizing when the optimal safety conditions hold except in the cases where regulations dictate the number.

The calculation for the sizing of screwed joints is shown in Chapter 7, together with the other elements needed to understand it.

## 6.6. Plates

Various types of plates are used in seismic reinforcement. They vary according to function.

Bent plates are one of the types most used in seismic reinforcement, where an elbow-shaped plate supports the wooden flooring joists, at the same time making a connection between the walls below the floor.

Figure 30 - Bent plate (Source: M. M. T. E. C.).

The plate is bolted to the walls, while the beams can be simply supported by the walls or via screwed joints.

Another type used are plates that reinforce existing beams, normally of wood, through the application of steel plates fastened to the beams thus making beams of mixed wood/steel composition (Appleton, 2003).

The proper functioning of this method must be attentive to the relationship between the elasticity of steel and wood in order to homogenise the composed section, choosing the height and thickness of the plate as a function of this (the coefficient of homogenisation should be 20:1) (Appleton, 2003).

Besides these two most common methods, there are various ways to use sheet metal, it is possible to create plates appropriate to each situation.

Figure 31 - Reinforcement of wooden beams with bolted steel plates (Source: M. M. T. E. C.).

## 6.7. Girders

Girders are use to replace elements that are in poor condition or to reinforce weakened zones of buildings. Its functions are very similar to those of girders used in new construction.

In implementation, girders need to be anchored on the walls through metal plates and screwed joints or simply by relying on openings created in the walls where the ends of the girders are placed followed by concrete.

Figure 32 – Girder (Source: M. M. T. E. C.).

In order for this method to not be too intrusive, the final installation of the girder can be hidden with a false ceiling or even by isolating the girder with a moulding that will allow it to imitate the rest of the ceiling.

Figure 33 – Girder with a molding imitating the building (Source: Blasi et al, 1999).

## 6.8. Connections between Metal Components of the Roof Structure

Connections between the metallic components of the roof, while not normally a direct target in the seismic reinforcement of buildings, can promote better functioning as a unit of the whole building.

These connection are made through metal plates with screwed joints which permit the creation the joints of a truss. These fittings can be in different shapes (Figure 34), namely: single or chicken leg, cross, clamp or frame; these shapes all being used in different conditions. It is also possible to substitute some hangers for rods when necessary, structurally reinforcing the roof.

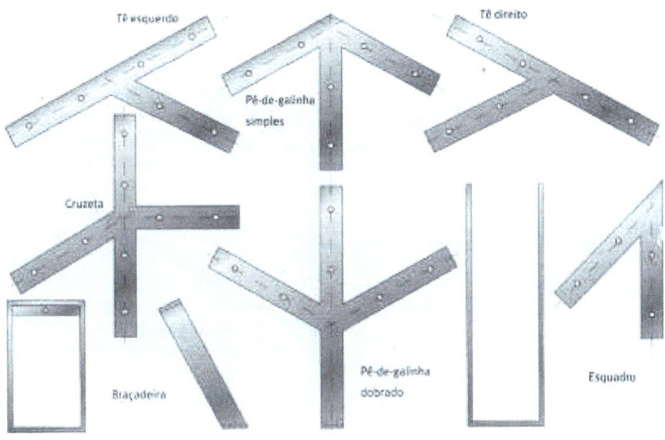

Figure 34 – Metallic Elements used in roof connections (Source: Appleton, 2004).

# Conclusion

A significant part of Portugal's architectural patrimony is constructed of natural stone masonry which confers an appreciable part of its historical value. It is therefore extremely important to maintain the character of buildings using methods that are unintrusive when retrofitting these buildings is justified.

The installation of metallic elements provides for unintrusive intervention and a higher level of success in seismic reinforcement. Despite the evidence of this type working well in Portugal being scanty, in Italy the effectiveness of these methods has been adequately demonstrated.

The application of metallic elements has various advantages, among them:

1   Transport;

2   Placement on the worksite;

3   Ease of operation in confined spaces;

4   Compatibility between form and function with regard to existing structures;

5   Reversibility and autonomy of these elements which facilitates maintenance and inspection of the structure;

6   Pre-fabrication.

Metallic elements are usually in stainless steel. This material due to its flexibility and high ductility is widely used in seismic reinforcement, because it allows it allows gradual rather than quick collapse in the event of structural failure; in other words, it gives warning of failure.

The lack of regulations for structural evaluation of old buildings makes necessary the use of regulations pertaining to new construction. Therefore, care must be taken in weighing values used to determine safety, given that there is a greater degree of uncertainty.

The accuracy of the structural evaluation will increase with the degree of detail of the inspection, and should include qualitative and quantitative tests.

To create a retrofitting project for each building the consideration of the following influences on the behaviour of structures is necessary:

1   The form;

2   The joints in the structure;

3   The building materials;

4   The imposed forces, accelerations and deformations.

It is important to develop a structural design, determine the damage, identify the characteristics of the materials and determine a course of action.

Having a perfect evaluation of the structure makes possible the determination of the best methods of intervention, which can be on three levels:

1   Safety;

2   Repair;

3   Reinforcement;

4   Restructuring.

Knowledge of the mechanisms of collapse whether they be of the first or second type brings a good understanding of the effects of earthquakes and the types of damage that can be caused to masonry buildings. With this, it is possible to determine the type of metallic element to be used in avoiding different types of damage.

Each component has a distinct role in strengthening the structural design. However, as mentioned earlier, the proper functioning of this network should be seen as a whole.

Several metallic elements are used in seismic reinforcement, rods, anchoring bars, plates, "cordolo" or crown belts, girders, metallic joints for roof components and screwed joints.

All of these metallic elements have distinct functions, as explained in this document, and are important in improving the response of the building to seismic action, specifically as responses to individual mechanisms of collapse.

# Bibliography

Aguiar, J., Appleton, J. e Cabrita, A., (2002). Guião de apoio à reabilitação de edifícios habitacionais. Lisboa, L.N.E.C..

Appleton, J., (2003). Reabilitação de edifícios antigos patologias e técnicas de intervenção. Amadora, Edições Orion.

Blasi, C., Borri, A., Di Pasquale, S., Malesani, P., Nigro, G., Parducci, A. e Tampone, G (1999). Manuale per la riabilitazione e la ricostruzione postsismica degli edifici: Regione dell'Umbria. Roma, Dei tipografia del genio civile.

Calado, L.. O aço na recuperação de edifícios. Disponível em http://www.dgpatr.pt [consulted in June of 2005]

Carocci, C., Ceradini, V., De Benedictis, R., Felice, G., Pugliano, A. e Zampilli, M (2000). Sicurezza e conservazione dei centri storici: Il caso Ortigia. Roma, Editori Laterza.

I.C.O.M.O.S.. Recomendações para a análise, conservação e restauro estrutural do património arquitectónico. Guimarães, University of the Minho.

Santos, S. Pompeu (2004). A reabilitação sísmica do património construído. In: Barros, J., Lourenço, P. e Oliveira, D.. Sísmica 2004: 6º Congresso Nacional de Sismologia e Engenharia Sísmica. Guimarães, University of the Minho, pp. 956-966.

www.ingramcontent.com/pod-product-compliance
Lightning Source LLC
Chambersburg PA
CBHW041205180526
45172CB00006B/1201